On a clear spring morning, a male song sparrow perches on top of a post and sings at the top of his voice. His song tells other birds, "This is the place for me! This is my territory! This is a place where I can raise a family!" You might hear his catchy song, too, because the place that he claims as his home might just be your own backyard—a yard, park, garden, or other green patch of Earth near you.

Your Backyard's Wild!

Every backyard, no matter how big or small, is a wildlife *habitat*—a place where wild creatures live and find everything they need to survive. That might seem like a lot to expect from your small piece of planet Earth. But any animal looking for a home really needs just a few basic things.

Every living creature needs **food**. Earthworms feed on soil and bits of decaying plant matter. Robins eat berries, insects, and earthworms. Almost any plant, insect, or other small creature might be food for another member of the natural community.

Many insects get all the **water** they need from their food, but some, including praying mantises and butterflies, need a refreshing drink from time to time. Most birds and mammals need to drink water. Many birds use water to bathe in, too.

Cover means places to hide and be safe. For example, a box turtle living in a garden may find cover in some flowers or tall grass. An animal also needs a safe place to raise its young. The mother box turtle lays her eggs in a nest that she digs out of soft soil or sand.

Which living things share your environment? How does your backyard meet their needs for food, water, cover, and a place to raise their young? The best way to find out is to become a backyard *naturalist*—a person who studies nature up close and firsthand. Who knows, you might even already be one!

Backyard Insects

Insects are the most numerous animals on Earth. In fact, scientists have counted over 88,000 different species (kinds) of insects in North America alone. Some of them are sure to show up in your backyard.

Beetles make up the largest group of insects. A bean beetle like this one might be found in a vegetable garden in your neighborhood. It eats the leaves of bean plants.

Leafhoppers also feed on plants. They have straw-like mouth parts that help them suck the juices of grasses, garden plants, wildflowers, bushes, and trees.

Scientists estimate that more than half of all insect species prey on other insects for food. This ladybug—which is also a kind of beetle—can gobble up dozens of little plant-eating insects called aphids (AY-fids) in a single day.

The praying mantis got its name from the way it often holds up its front legs as if it were praying. But you can also think of it as the *preying* mantis, because it eats nothing but other insects. Mosquitoes, flies, grasshoppers, beetles, and aphids are just a few of the many six-legged creatures that it hunts.

And if some colorful flowers catch your eye, chances are that you'll see insects like bees, wasps, butterflies, and ants hovering or crawling around. That's no accident. The insects and plants are partners.

A flower's bright colors and sweet scent send a special message to insects like this bumblebee—"Come and get your favorite foods! There's powdery pollen and liquid nectar here!" The busy bee accepts invitation after invitation, traveling from flower to flower. As it does, it leaves behind something that each flower needs—some grains of pollen from another flower of the same species. These pollen grains will help the flowers make new seeds.

Many flowers have special markings that act as "nectar guides" for their insect visitors. The dotted paths on this foxglove show an insect exactly where to go for a sip of nectar—and a good dusting of pollen, too.

Have you ever wondered why some flowers, such as tulips, morning glories, and daisies, close their petals at night and on rainy days? By closing up, they are able to save their nectar and pollen for a sunny day. That's when their insect pollinators are active.

Backyard Birds

Tap-tap-tap. Ya, ya, ya, ya. Cheerio, cheerily, cheery. Chip, chip, chip. The sights and sounds of birds are everywhere. If you pay close attention, you can learn about their different voices, sizes, shapes, colors, and ways of life. You can also learn about your backyard by looking at it from the birds' point of view. To them, it's a place to eat, rest, hide, and maybe even raise a family.

Woodpeckers use their strong beaks to chisel nest holes in dead trees. This red-headed woodpecker is cleaning house, emptying out most of the wood chips that fell in during excavation. It will be sure to leave some inside, though—they'll make a cozy lining for the nest.

Like many birds, white-breasted nuthatches eat a variety of foods. These little birds get their name from their skill at cracking open acorns and other hard nuts. Some people also call them "upside-down birds" because they walk *down* tree trunks as they look for insects to eat.

Robins and most other backyard birds build nests in the branches of trees and shrubs. The trees and shrubs also provide the birds with cover from predators and harsh weather. And some, like this firethorn bush, offer a bonus of fruit to eat.

Birds that spend a lot of time on the ground pecking for seeds and insects are known as *ground feeders*. This dark-eyed junco is one. Sparrows, cardinals, blue jays, and mourning doves are some other familiar ground feeders.

In backyards across North America, millions of people fill and put out bird feeders. The feeders range from homemade models like this pinecone dabbed with peanut butter to clear plastic bubbles that look like seed-carrying flying saucers. Why do they do it?

When people put food out, they may help birds survive, especially in places that have icy or snowy winters. Providing food is also a great way to bring birds—like this little white-crowned sparrow—into closer view. Backyard bird watchers soon start discovering answers to such questions as: Which birds live in my area? What do they eat? What gives each species its own special "personality"?

Blue jays like this one live east of the Rocky Mountains. They love peanuts and sometimes stash them under fallen leaves and in other secret hiding places.

Cardinals live mostly east of the Rockies. Pairs of bright red males and rosy-brown females often investigate and visit feeders together.

Black-capped chickadees are almost always on the go. In the winter, they zip from tree to tree in search of tasty insect eggs. And in most parts of the country, they're often the first eager customers at a freshly-filled feeder.

Goldfinches can be spotted throughout most of the United States. They approach feeders with a rising and dipping style of flight that reminds many people of a roller coaster.

Mourning doves are found in every state except Hawaii. They bob their heads as they slowly strut along the ground, often picking up seeds that other birds have spilled. Backyard bird watchers also know them by their soft, *coo-coo-cooing* song and the whistling sound their wings make as they take off in flight.

They're Everywhere!

Are you used to seeing one or more of these animals? If you are, that's not surprising. They all thrive in cities, suburbs, towns, and other places built and used by people. And they do so because they've learned to take advantage of the special opportunities that come with the territory.

Gray squirrels don't mind having people around. In fact, they often prefer things that way. As *omnivores*—animals that eat a wide range of foods—their wild diet includes acorns, berries, tree buds, and insects. Birdseed, apples, and cookies are just a few of the items they add to their menu when they make themselves at home near people.

You can't call a crow a picky eater, either. Scientists have found that these birds eat over 600 different foods!

Raccoons and crows are two more bold and clever omnivores. Like squirrels, they've learned that where there are people, there's bound to be food.

Raccoons make their rounds at night, looking for anything from leftover dog food to fresh vegetables from the garden.

Both house sparrows and starlings prefer to live near people. They come not only for the food but also for the shelter that human habitats offer. In the spring, you might find them building their nests under the roofs of houses. Both birds also nest in tree cavities and abandoned woodpecker holes, often driving away other species who use the same kind of nesting sites.

Tracks and Traces

Backyard wildlife isn't always easy to spot. Some animals are shy and cautious as they move about. Some stay hidden much of the time. Some are tiny. If you look carefully, though, you might find clues about the whereabouts and activities of these secretive creatures.

Snow-covered yards or fields are logical places to look for animal tracks. So are the muddy edges of streams and rivers. You may not always be able to tell exactly whose trail you're on, but you can put certain pieces of the puzzle together by observing closely. Are the animal's feet tiny, small, medium, or large? Does it seem to walk, run, or hop? Does the trail begin or end at the base of a tree? These tracks were made by an opossum, a mammal that is active at night.

Holes of various sizes might prove to be the work of *burrowing* animals—animals that dig and take shelter underground. Chipmunks, woodchucks, and skunks often burrow in backyards and nearby woodlands. This neatly-dug little hole belongs to a chipmunk.

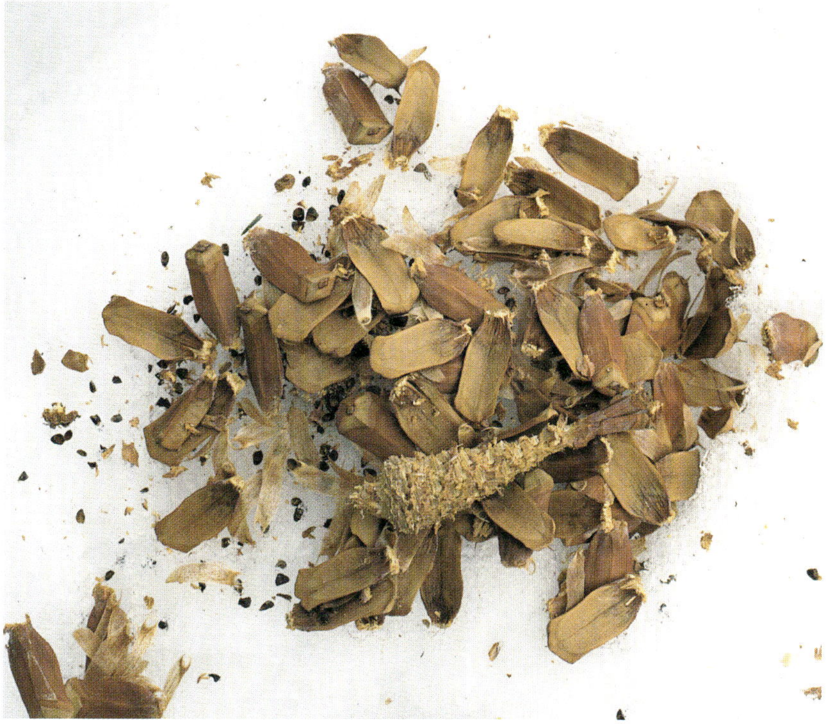

Animals that eat the nuts inside acorns or pinecones often leave behind little scrap heaps known as *midden piles*. A midden pile might be a sign of gray squirrels. But it's even more likely to be a sign of harder-to-spot species such as red squirrels or flying squirrels. This pile of pinecone scales is the work of a red squirrel.

What's going on here? These squiggly lines were made by tiny insects known as *leaf miners*. The leaf miners are actually the larvae (LAHR-vee)—or young—of certain species of moths, beetles, and flies. They find both food and protection as they tunnel their way through the insides of leaves.

Many species of web-building spiders stay out of sight, but they're usually not far from their webs. The purpose of any spider web—including this funnel-shaped web made by a grass spider—is to trap insects. So webs are a sure sign that there are insects around as well.

Welcoming Wildlife

Backyards, schoolyards, parks, and gardens are places where people and wildlife meet. Lots of people are discovering ways to make these places even more inviting for wild creatures of all kinds. Look around your own community. You might see some of these wildlife-friendly ideas catching on.

The grounds in front of this apartment house in Washington, D.C., show that a yard doesn't have to be big to roll out the welcome mat for birds and other animals. As long as there is a variety of plants of different sizes, a variety of creatures will find food and cover.

In many areas, bluebirds like this one have lost their homes. So have many other birds, including woodpeckers, nuthatches, and chickadees. That's because the tree holes they normally nest in disappear as dead trees are cut down, or because more aggressive species like house sparrows and starlings drive them away. Many people are helping out by putting up nesting boxes.

Gardeners are choosing flowers that suit the tastes of birds and butterflies. This tall sunflower, for instance, offers nectar to a monarch butterfly as it blooms.

After the petals fall off and the seeds mature, it becomes a home-grown feeder for nuthatches and lots of other birds.

No backyard habitat is complete without a source of fresh water. A birdbath perched on a tree stump, a post, or even hung from a branch gives chickadees and other birds a place to drink and bathe.

A shallow container placed right on the ground makes a great watering hole for squirrels and other small mammals, too.

Whether your backyard is a big city park, a small country garden, or a grassy lot outside your school, it's your part of the natural world. It's yours to explore, and it's close to home. Your backyard is quite a place.